By
Charis Mather

BookLife PUBLISHING

©2023
BookLife Publishing Ltd.
King's Lynn, Norfolk
PE30 4LS, UK

All rights reserved.
Printed in Poland.

A catalogue record for this book is available from the British Library.

ISBN: 978-1-80155-883-9

Written by:
Charis Mather

Edited by:
William Anthony

Designed by:
Drue Rintoul

All facts, statistics, web addresses and URLs in this book were verified as valid and accurate at time of writing. No responsibility for any changes to external websites or references can be accepted by either the author or publisher.

AN INTRODUCTION TO BOOKLIFE RAPID READERS...

Packed full of gripping topics and twisted tales, BookLife Rapid Readers are perfect for older children looking to propel their reading up to top speed. With three levels based on our planet's fastest animals, children will be able to find the perfect point from which to accelerate their reading journey. From the spooky to the silly, these roaring reads will turn every child at every reading level into a prolific page-turner!

CHEETAH
The fastest animals on land, cheetahs will be taking their first strides as they race to top speed.

MARLIN
The fastest animals under water, marlins will be blasting through their journey.

FALCON
The fastest animals in the air, falcons will be flying at top speed as they tear through the skies.

Image Credits

All images courtesy of Shutterstock.com. With thanks to Getty Images, Thinkstock Photo and iStockphoto. Cover and throughout – Vadim Sadovski, Ortodox, Triff, pne, M.Aurelius, lassedesignen. p4–5 – mooremedia, zhengzaishuru. p6–7 – max dallocco, Pixus. p8–9 – Negro Elkha, somavarapu madhavi. p10–11 – cigdem, 3000ad. p12–13 – Mirai, nasti23033. p14–15 – WP, CC BY-SA 3.0, via Wikimedia Commons, Vadim Sadovski, Jurik Peter. p16–17 – sripfoto, AstroStar. p18–19 – EHT Collaboration, CC BY 4.0, via Wikimedia Commons, jivacore. p20–21 – EduardoMSNeves, CC BY-SA 4.0, via Wikimedia Commons. p22–23 – Dima Zel, Dotted Yeti.

Contents

PAGE 4   Super Space
PAGE 6   The Sun
PAGE 8   The Earth
PAGE 10   Planets
PAGE 12   Moons
PAGE 14   Our Solar System
PAGE 16   Stunning Stars
PAGE 18   Grand Galaxies
PAGE 20   Magnificent Milky Way
PAGE 22   More to Explore
PAGE 24   Glossary and Index

Words that look like this can be found in the glossary on page 24.

SUPER Space

We can see space above us every night. However, there's still a lot that we do not know about it.

We do know that space is full of stars, planets, moons and more.

Let's learn more about our amazing universe.

THE Sun

The Sun is enormous. It would take more than 1 million Earths to fill up the Sun.

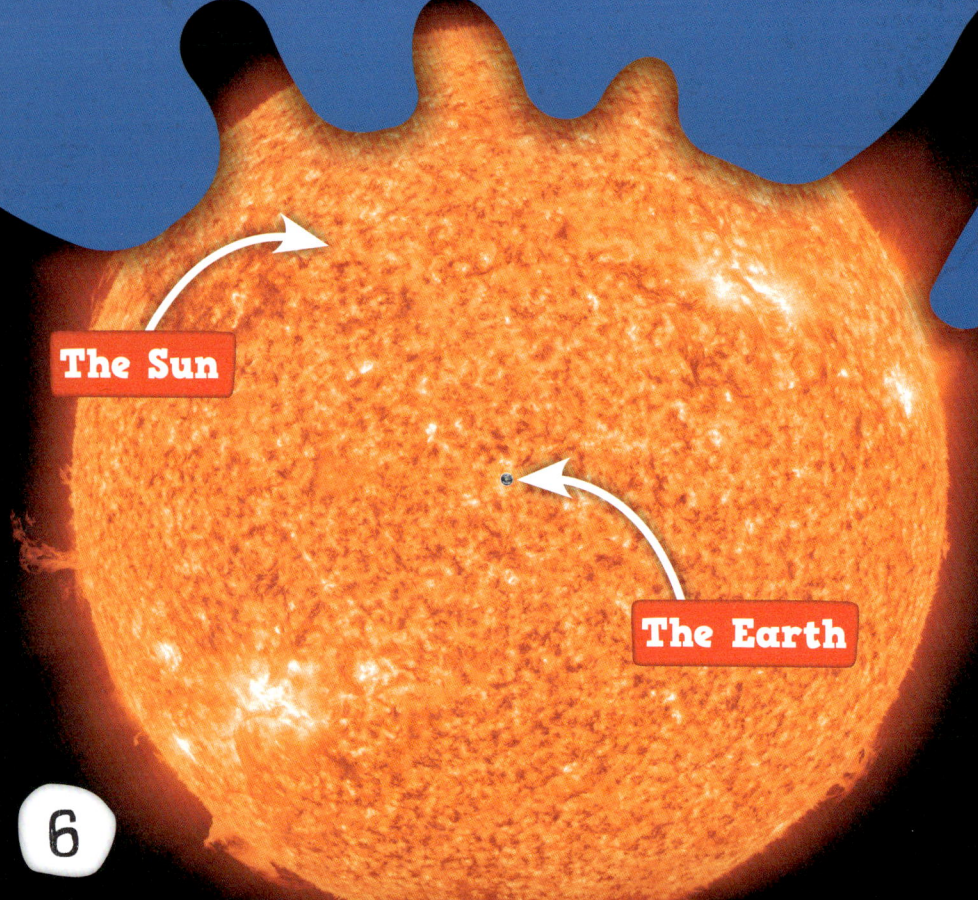

The Sun

The Earth

The Sun is also very hot. The centre of the Sun is around 15 million <u>degrees Celsius</u>.

That's hot!

THE Earth

We live on Earth.

Earth is very different from all the other places we know about in space.

Earth is full of life. It has lots of water and <u>oxygen</u>. Humans and animals need these to live.

Planets

Planets are large objects in space that travel around a star.

Earth is one of eight planets going around the Sun.

Planets are different sizes and colours.

Saturn

Jupiter, Saturn, Uranus and Neptune are planets that have rings of rocks floating around them.

moons

Moons travel around planets just like planets travel around stars. Some planets have lots of moons.

Some of Jupiter's moons

The Earth has a moon. Sometimes, the Moon looks like it is a different shape because of the way the Sun's light hits it.

OUR Solar System

Everything that goes around the Sun is part of something called the Solar System.

The Solar System has eight planets and over 140 moons.

It also includes rocks that are smaller than planets.

STUNNING Stars

The night sky is full of stars.

There are billions and billions of them in the universe.

Stars sometimes look like they are twinkling. This is because of the way that light comes through Earth's <u>atmosphere</u>.

GRAND Galaxies

Stars can group together to make galaxies. They can form different shapes.

Spiral

Elliptical

Irregular

18

Scientists think that galaxies have <u>black holes</u> at their centres that can pull things towards them. Our own galaxy has one.

A black hole

MAGNIFICENT Milky Way

We are in a galaxy called the Milky Way. We are 26,000 <u>light-years</u> from the centre.

The Sun

The Milky Way is a spiral galaxy. Sometimes, you can see part of it from Earth.

MORE TO Explore

Space is huge.

You could spend your whole life travelling through space. You would never reach the end.

We've only seen a tiny part of our universe.

There will always be more to explore!

Glossary

atmosphere — the mixture of gases that make up the air and surround the Earth

black holes — areas in space with a strong force that can pull objects and light into them

degrees Celsius — something that scientists use to describe how hot or cold something is

light-years — the amount of time that it takes for light to travel in one year

oxygen — something that living things need in order to survive

Index

life 9, 22
light 13, 17, 20
night 4, 16
rings 11

rocks 11, 15
scientists 19
shapes 13, 18
sizes 11

universe 5, 16, 23